YOUR KNOWLEDGE HAS VALUE

- We will publish your bachelor's and master's thesis, essays and papers

- Your own eBook and book - sold worldwide in all relevant shops

- Earn money with each sale

Upload your text at www.GRIN.com
and publish for free

Purvesh Shah

Organic preparation based on name reaction

GRIN Verlag

Bibliografische Information der Deutschen Nationalbibliothek:

Die Deutsche Bibliothek verzeichnet diese Publikation in der Deutschen National-
bibliografie; detaillierte bibliografische Daten sind im Internet über http://dnb.d-
nb.de/ abrufbar.

Imprint:

Copyright © 2013 GRIN Verlag GmbH
Druck und Bindung: Books on Demand GmbH, Norderstedt Germany
ISBN: 978-3-656-66077-4

This book at GRIN:

http://www.grin.com/en/e-book/272163/organic-preparation-based-on-name-
reaction

GRIN - Your knowledge has value

Der GRIN Verlag publiziert seit 1998 wissenschaftliche Arbeiten von Studenten, Hochschullehrern und anderen Akademikern als eBook und gedrucktes Buch. Die Verlagswebsite www.grin.com ist die ideale Plattform zur Veröffentlichung von Hausarbeiten, Abschlussarbeiten, wissenschaftlichen Aufsätzen, Dissertationen und Fachbüchern.

Visit us on the internet:

http://www.grin.com/

http://www.facebook.com/grincom

http://www.twitter.com/grin_com

Organic Preparation based on some name reactions

Dr. Purvesh J. Shah

Organic Chemistry Practicals

Practical-1: Preparation of p-Iodo nitrobenzene from p-Nitroaniline.

Diazotization:

Diazonium salts were prepared first by Peter Griess in 1858 by the action of nitrous acid on the salts of an aromatic amine.

Most of Diazonium salts are unstable at room temperature. So reaction is carried out at 0°c and solution used immediately.

Coupling takes place only in weakly acidic or alkaline solution. In strongly acid solution the hydroxyl group is undissociated and the amino group present as its salts.

Coupling takes place in the unhindered para position of the coupler. If the para position occupied coupling takes place at ortho position. If both positions are occupied coupling not takes place occasionally the group in the para position is displaced.

The process of converting an amine into the Diazonium salt is called diazotization. The dry diazonium salt is unstable and explosive. Aryldiazonium salts are highly reactive compounds and undergo many useful reactions. They have proved to be valuable synthetic intermediates for the preparation of a fairly wide range of aromatic derivatives.

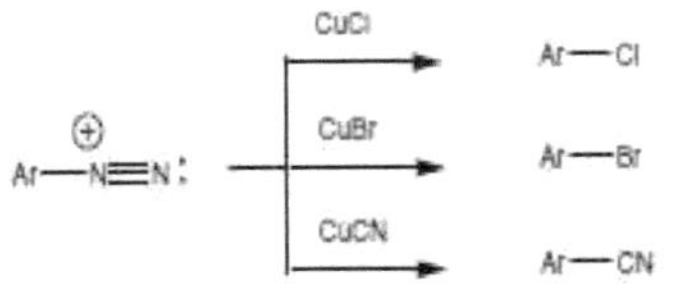

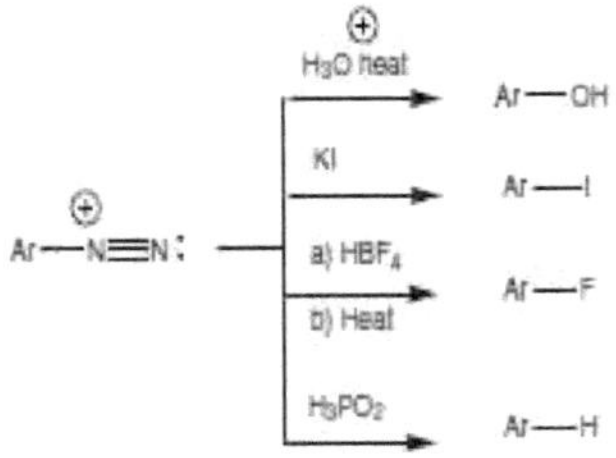

Organic Chemistry Practicals

Reaction Scheme:

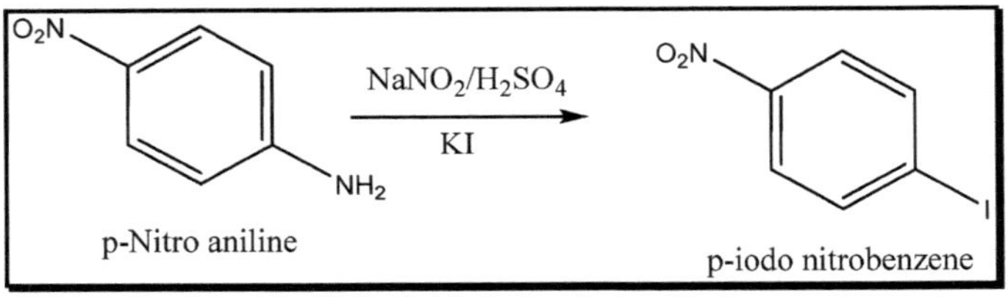

p-Nitro aniline → p-iodo nitrobenzene

Requirements:

p-nitro aniline : 1g

Conc.H2SO4 : 1 ml

0% NaNO2 solution : 5 ml

20% KI solution: 10 ml

Procedure:

Mix p-nitro aniline and concentrated sulphuric acid thoroughly to Make-it into a paste in a 50 ml beaker. Add 10 ml Water to it and stir to dissolve p-nitro aniline completely, (Heat gently, if required). Cool the solution to 0°C in an ice bath and diazotize by adding the solution of sodium nitrite drop by drop to it with constant stirring. The temperature should be controlled below 5°C throughout the experiment. Take potassium iodide solution in a 250 ml beaker and cool it to 5°C. Drop wise add diazonium salt solution to potassium iodide solution with vigorous stirring. The reaction proceeds with evolution of N_2 (Frothing occurs). After the addition is over, allow the reaction mixture to stand at room temperature for 30 minutes. Collect the precipitated product by filtration, wash with water and crystallise from **ethanol**.

Melting Point: crystallized form melts at 171°C.

Practical-2: Preparation of Dibenzalacetone from Benzaldehyde. [Ciaisen-Schmidt Reaction]

Ciaisen-Schmidt Reaction:

The condensation of aromatic aldehydes having no α-hydrogen with aliphatic aldehyde, ketone or ester having active hydrogen in presence of alkali solution give α, β-unsaturated aldehyde or ketone is known as claisen-Schmidt reaction.

By using this reaction we can synthesis many natural products, perfumery compound.

Reaction Scheme:

Benzaldehyde

Acetone/NaOH

Dibenzalacetone

Requirements:

Benzaldehyde	: 2.0 ml	Acetone: 0.8 ml
10% NaOH solution	: 4.0 ml	R-Spirit: 20 ml

Procedure:

Dissolve benzaldehyde and pure acetone in 20-ml of R-Spirit contained in a conical flask of 100 ml capacity. Add 4 ml of 10% aqueous sodium hydroxide alkali dropwise to the former solution over a period of 30 minutes. Shake the mixture vigorously in the securely corked flask for about 10 minutes (releasing the pressure from time to time if necessary). Then allow standing for 30 minutes with occasional shaking. Finally cool in ice water for a few minutes. During the shaking, the dibenzal acetone separates first as a fine emulsion which then rapidly forms pale yellow crystals. Filter at the pump, wash well with water to eliminate traces of alkali and then drain thoroughly. Recrystallise from hot R-Spirit. The dibenzalacetone is obtained as pale yellow crystals.

Melting Point: crude product melts at 106°C

Purified product melts at 112°C.

Practical:3 Preparation of Benzalnilide from Benzophenone.

[BECKMANN REARRANGEMENT]

BECKMANN REARRANGEMENT:

The Beckmann rearrangement consists of ketoximes into substituted amides by heating with some acidic reagent.

The rearrangement is stereo specific and the group that normally migrates is one that is anti with respect to hydroxyl. This method is often used to determine the configution of oxime.

Reaction Scheme:

Requirements:

Step-1 Benzophenone oxime :

Benzophenone: 1 g   Hydroxylamine hydrochloride: 0.6 g

Sodium hydroxide: 1g   R-Spirit : 8ml

Dilute HCl: 6 ml con.HCl in 30 ml water

Step-II Benzophenone oxime:

Benzophenone oxime : 1 g

Anhydrous ether   : 10 ml

Conc. H_2SO_4 : 1.6 ml

Procedure:

(i) Benzophenone oxime from Benzophenone

Dissolve benzophenone and hydroxylamine hydrochloride in 5 ml 80% aqueous R-Spirit in a 50 ml round bottom flask equipped with a reflux condenser. Continuously stir the solution by swirling the flask. Prepare a solution of sodium hydroxide in 1 ml water. Portion wise add this solution to the former solution. If the reaction becomes vigorous, cool-the flask under running tap water. After complete addition, reflux the reaction mixture for 30 minute. Allow the mixture to cool and dilute with 10 ml Water. Separate the unreacted benzophenone by Filtration (if found). Cool the filterate and pour it into dilute hydrochloric acid with constant stirring. Filter the precipitated benzophenone oxime, wash with cold water and crystallise from methanol.

Melting Point: 142°C.

N.B.: *Oxime should be preserved in vacuum desiccator filled with pure carbon dioxide to prevent its decomposition by oxygen and traces of moisture.*

(ii) Benzanilide form Benzophenone oxime

Dissolve benzophenone oxime in anhydrous ether in a distilling flask equipped completely for distillation. Add concentrated H_2SO_4 to the ethereal solution. Shake it thoroughly to ensure proper mixing. Consciously distill out solvent and, other volatile products on a boiling water bath. Trap the distillate in a 250 ml beaker containing tap eater. Cool the flask. Pour 20-25 ml cold water into it with stirring. Boil it for several minutes. Break-up any lumps which may form during boiling. Cool it and decant the supernant liquid. Crystallization may be performed in the same flask using.-boiling R-Spirit.

Melting Point: 163°C in the purified form.

Practical: 4 Preparation of 1-Phenylazo-2-naphthol from β-Naphthol. [Diazotization and Coupling]

Reaction Scheme:

1-Phenylazo-2-naphthol

Requirements:

Aniline: 1 ml	Conc. HCl: 3ml
10% NaNO$_2$ solution: 8ml	β –naphthol: 1.5 gm
10% NaOH: 10 ml	Acetic acid, water, Ice.

Procedure:

Dissolve aniline in concentrated HCl in a 50 ml beaker. Add 4 ml water solution cool to it 0-5°C. Diazotize the cold solution of aniline hydrochloride by drop wise addition of sodium nitrite solution. Prepare a solution of β -naphthol in 10% aqueous NaOH in a 100 ml. beaker. Cool the solution to 5°C in an ice-salt bath. Slowly add the diazonium salt solution to the naphthol solution with constant stirring. Red precipitate of the dye will separate. After the addition is complete, allow the reaction mixture to stir for 30 minutes in the ice bath with occasional stirring. Filter through a Buchner funnel water. Crystallize the product from acetic acid (5 ml). Filter by suction filtration, wash with little alcohol to eliminate traces of acetic acid and dry.

Melting Point: 134°C in the purified form.

Practical: 5 Preparation of 9,10-dihydroanthracene-α,β-succinic anhydride from Anthracene.[DIELS-ALDER REACTION]

DIELS-ALDER REACTION:

The reaction between diene with dienophile is called Diels-alder reaction. The neat result is formation of 2 new sigma bonds and 1 new pi-bond. It is 4+2 cycloaddition.

A method for planning synthesis that involves reasoning backward from the target molecule through various levels of precursor's and thus finally to the starting materials is called retro synthesis. It is a reverse of the chemical reaction.

Reaction Scheme:

Anthracene

Maleic anhydride

9,10-dihydroanthracene-α,β-succinic anhydride

Requirements:

Anthracene : 0.50 gm   Maleic anhydride: 0.25 gm

Dry p-xylene: 10 ml   Activated charcoal

Procedure:

Place anthracene, maleic anhydride and p-xylene in a 50ml round bottom flask fitted with a reflux condenser. Boil the mixture under reflux for 20 minute shaking the flask frequently during first 10 minute. Allow to cool to room temperature, add 0.25 g decolorizing carbon and boil for further 10 Minute. Filter the hot solution through a small-preheated Buchner funnel. Collect the solid from the filtrate which separates upon cooling. Dry the product in a vacuum desiccator containing paraffin wax shavings to absorb traces of p-xylene.

Melting Point: The purified product melts at 262-264°C.

N.B.: *The exposure to air tends to cause hydration of the anhydride portion of the product. The product should be stored in a well stopperred glass bottle (sample tube).*

Practical: 6 SANDMEYER REACTION of Aromatic amines

[Aniline]

Reaction Scheme:

Requirements:

Copper sulphate : 1.0 gm	Aromatic amines [Aniline]	: 0.5 g
Sodium chloride : 0.9 gm	10% aq. $NaNO_2$ solution	: 3 ml
Copper foil : 0.5 g	Conc.HCI.	Water

Procedure:

Take copper sulphate, sodium chloride and copper foil in a 50ml round bottom flask. Add 5 ml water and 3 ml of concentrated HCl to it. Reflux the content till colorless solution is obtained. Cool this solution to 0-5°C in an ice bath. After transferring it into a 250ml beaker.

Meanwhile, prepare a uniform paste of given **aromatic amine** in 1ml concentrated HCl in a 50 ml beaker. Add 5ml water and heat gently to get clear solution. Cool this solution to 0°c in ice bath. Diazotize the cooled solution by drop wise addition of 10% aqueous sodium nitrite solution. The mass should be stirred continuously by maintaining the temperature below 5°C. This cold diazonium salt solution is to be slowly added to the CuCl solution kept in 250ml beaker. The reaction Proceeds with frothing. The diazonium salt solution should be added in such a controlled way that the reaction mass does not come out of the beaker through frothing. After the addition is complete, allow the reaction mixture to stand at room temperature for 1hour with frequent stirring. Filter the crude product, wash with cold water and crystallize from proper solvent.

Practical: 7 Preparation of 7-hydroxy-4-methylcoumarin from Resorcinol. [PECHMANN CONDENSATION]

PECHMANN CONDENSATION:

The reaction of phenol with β-keto ester in presence of acid to give coumarins is known as Pechmann condensation.

Pechmann reaction finds widely use in synthesis of coumarins and cromones. Coumarins derivatives used as anticoagulating agents, in dyes as fluorescent brighting agents, flavoring agents, medicines.

Reaction Scheme:

Requirements:

Resorcinol   : 0.5gm

Ethylacetoacetate : 0.5 ml

Cont. Sulphuric acid: 2 ml

Procedure:

Take concentrated sulphuric acid in a 50 ml r.b. flask and cool it to 5⁰C by swirling the flask in an ice-bath. Prepare a solution of resorcinol in ethylacetoacetate in a test tube. Add this solution drop by drop to the cold sulphuric acid by maintaining the temperature of the mixture well below 10°C.Continue stirring for 30 minutes. Pour the reaction mixture on to ~25 gm of crushed ice. Fitter the separated product and wash several times with water. Dissolve the product in cold 10% aqueous NaOH solution. Reprecipitate 7-hydroxy-4-methylcoumarin by adding 10% aqueous HCl till the solution becomes acidic to litmus paper. Filler the reprecipitated product, wash with water and crystallize from ethanol; using activated charcoal if necessary.

Melting Point: The purified product melts at 185°C.

Practical: 8 Preparation of 1,2,3,4-Tetrahydrocarbazole from Cyclohexanone. [FISCHER INDOLE SYNTHESIS]

FISCHER INDOLE SYNTHESIS:

The formation of Indole derivative by heating aryl hydrazones of suitable aldehyde, ketone or ketonic acid in presence of an acid catalyzed (zinc chloride, poly phospheric acid, H_2SO_4) is known as Fischer Indole synthesis.

It can be used in the synthesis of many heterocyclic compounds like carbazoles indole-3-aceticacid,3-methyl oxindole, tryphamide.

Reaction Scheme:

Requirements:

Cyclohexanone   : 0.5 ml

Acetic acid   : 5 ml

Phenyl hydrazine : 0.5 ml

Procedure:

Mix cyclohexanone and acetic acid in a 50 ml round bottom flask. Carefully add phenyl hydrazine to it with constant shaking. Heat the mixture under reflux for 30 minutes. Cool the flask, filteroff the brown solid, wash thrice with cold Water taking the aliquot of 10 ml each time. Crystallize from 50%

aqueous methanol using charcoal.

Melting Point: The product should melt at 116-117°C.

Caution: *Phenyl hydrazine neither should be inhaled nor should be allowed to contact with the skin. It is **POISONOUS.***

Practical: 9 Preparation of Iodoform from Acetone. [Haloform Reaction]

Haloform Reaction:

Methyl ketones, as well as acetaldehyde, are cleaved into a carboxylate anion and a trihalomethane (a haloform) by the *Haloform reaction.* The respective halogen can be chlorine, bromine or iodine. The methyl group of a methyl ketone is converted into an α,α,α-trihalomethyl group by three subsequent analogous halogenation steps, that involve formation of an intermediate enolate anion by deprotonation in alkaline solution, and introduction of one halogen atom in each step by reaction with the halogen.

A halogen substituent to the carbonyl group makes adjacent hydrogen more acidic, and further halogenation will take place at the same carbon center. The α,α,α-trihaloketone can further react with the hydroxide present in the reaction mixture. The hydroxide anion adds as a nucleophile to the carbonyl carbon; the tetravalent intermediate suffers a carbon–carbon bond cleavage The reaction also works with primary and secondary methyl carbinols. Those starting materials are first oxidized under the reaction conditions to the corresponding carbonyl compound.

Its synthetic importance, the haloform reaction is also used to test for the presence of a methylketone function or a methylcarbinol function in a molecule. Such compounds will upon treatment with iodine and an alkali hydroxide lead to formation of iodoform (*iodoform test).* The iodoform is easily identified by its yellow colour, its characteristic odour and the melting point.

Reaction Scheme:

Requirements:

Acetone   :1 ml

10% aqueous NaOH: 4 ml

Iodine solution :(10g iodine dissolved in a solution of 20g KI in 100ml water)

Procedure:

Take acetone in a 50 ml r.b. flask. Mix it with 4 ml 10% aqueous NaOH solution and 10 ml water. Dropwise add iodine solution with continuous shaking till the colour of iodine persists. Heat the mixture in a waterbath. maintained at 60°C. If color of iodine disappears, add little excess of iodine solution. Heat the mixture in a water-bath to settle down the yellow precipitate Iodoform. Filter the product and crystallize from 1:1 aqueous methanol.

Melting Point: The product should melt at 119°C.

N.B.: The Haloform Reactions is known as Lieben Iodoform Reaction. The reaction is also possible with acetaldehyde, ethanol and secondary methylcarbinols along with methyl ketones.

Practical: 10 Preparation of 2-Hydroxy-4-methylquinoline from Aceto acetanilide. [Knorr Quinoline Synthesis]

Knorr Quinoline Synthesis :

Quinoline is a bicyclic heterocyclic having a benzene ring fused with a pyridine ring in 2,3 –positions. It occurs in coal-tar along with other bases and can be isolated from it.

A mixture of aniline, glycerol and sulphuric acid is heated in presence of a mild oxidizing agent such as nitrobenzene. The reaction being exothermic and $FeSO_4$ is also added as moderator. This reaction is known as skraup synthesis.

Quinoline is used as a high boiling basic solvent in organic reaction, in manufacture of pharmaceuticals, dyes, and insecticides. It gives electrophilic substitution reaction like benzene and forms salts with acids like pyridine.

Reaction Scheme:

Acetoacetanilide

conc.H_2SO_4

2-Hydroxy-4-methylquinoline

Requirements:

Acetoacetanilide   : 1g   conc.H_2SO_4: 2 ml

Organic Chemistry Practicals

Procedure:

Take 2 ml conc.H_2SO_4 in a 50 ml round bottom flask. Place a thermometer in it and heat the acid to 70°C in a boiling water bath. Add 1g of acetoacetanilide in small portions by maintaining the temperature of the mixture at 70°C by intermittent cooling. Heat the reaction mixture for 60 minutes in a boiling water bath. Cool and add 25 ml water to it with vigorous shaking. Filter the product; wash with water (3 X 5 ml). Crystallize from hot methanol.

Melting Point: The product should melt at 223-224°C.

Practical: 11 Preparation of 2,4-Dihydroxybenzoicacid from Resorcinol.[Kolbe - Schmitt Reaction]

Kolbe - Schmitt Reaction:

Carboxylation of phenolates/synthesis of salicylic acid.

Carbon dioxide reacts with phenolates to yield salicylate; with less reactive mono-phenolates, the application of high pressure may be necessary in order to obtain high yields. This reaction, which is of importance for the large scale synthesis of salicylic acid, is called the *Kolbe–Schmitt reaction.*

The Kolbe–Schmitt reaction is limited to phenol, substituted phenols and certain heteroaromatics. The classical procedure is carried out by application of high pressure using carbon dioxide without solvent; yields are often only moderate. In contrast to the minor importance on laboratory scale, the large scale process for the synthesis of salicylic acid is of great importance in the pharmaceutical industry.

Reaction Scheme:

Requirements:

Resorcinol : 1 g

$KHCO_3$: 12g

Procedure:

Take resorcinol and 20 ml water in a 50 ml round bottom flask fitted with a reflux condenser. Heat the mixture on sand bath for an hour. During reflux, add potassium bicarbonate to the boiling mixture at an interval of 5minute taking about 1 g of the bicarbonate each time. Acidify the hot solution by adding 3 ml concentrated HCl from a dropping funnel with its stem touching to the bottom of the flask. Allow the content to cool down to room temperature and then cool in ice bath: Resorcylic acid gets separated, filter and crystallize it from hot water.

Melting Point: The reported melting point is 216-217°C.

Practical: 12 Preparation of Benzyl alcohol and Benzoic acid from Benzaldehyde. [Cannizzaro Reaction].

Cannizzaro Reaction:

The reaction in which two aldehyde groups are transformed into the corresponding hydroxyl and carboxyl functions, existing separately or in combination as an ester, has been termed the Cannizzaro reaction.

Reaction Scheme:

| Benzaldehyde | Benzyl alcohol | Benzoic acid |

Requirements:
Benzaldehyde: 1.0 ml   KOH (Pellets): 0.625 gm

Procedure:

Dissolve potassium hydroxide in 5.0 ml of water in 100 ml round bottom flask. Add freshly distilled benzaldehyde and cool the mixture. Add few drops of water to just dissolve ppt. of potassium benzoate and shake the mixture for 20 minutes. Extract the mixture with three, five ml portions of ether and collect combined ether extract in another flask. Cool the remained alkaline solution acidified with 5.0 ml of HC1 (1:1). Again cool and filter the benzoic acid. Crystallize benzoic acid from hot water as colorless needles. The combined ether extract is dried over magnesium sulphate and distilled. The residue containing benzyl alcohol is distilled using flame. Cool the alkaline solution.

Practical: 13 Preparation of Chalcon from acetophenone.

[Aldol Condensation].

Aldol Condensation:

Reaction of aldehydes or ketones to give β -hydroxy carbonyl compounds.

The addition of the β-carbon of an enolizable aldehyde or ketone to the carbonyl group of a second aldehyde or ketone is called the *aldol reaction*.

It is a versatile method for the formation of carbon–carbon bonds, and is frequently used in organic chemistry. The initial reaction product is a β-hydroxy aldehyde (aldol) or β-hydroxy ketone (ketol) **3**. A subsequent dehydration step can follow, to yield an α,β-unsaturated carbonyl compound **4**. In that case the entire process is also called *aldol condensation*.

Reaction Scheme:

Requirements:

NaOH: 0.5g/2ml   R-spirit: 2.3ml

Acetophenone :0.5g   Benzaldehyde:0.8g

Procedure:

Place a solution of NaOH dissolved in 2 ml. water and R-spirit in a 100 ml flask. Cool the solution with ice and add freshly distilled acetophenone. Shake the mixture and add pure benzaldehyde. Keep the mixture at 25°C and stir until mixture is so stick that stirring no longer is effective.(1.5 hour).Pour the content in ice, shake and keep for some times. Filter the product with suction Buchner funnel. Wash with cold water until the washings are neutral to litmus paper, then with ice-cold R-spirit. Dry the product in air weigh product and recrystallize from R-spirit Warm to 50°C.

Practical: 14 Preparation of β-dimethyl amino propinphenone hydrochloride from acetophenone. [Mannich Reaction].

Mannich Reaction:

The Mannich reaction involves the condensation of ammonia or a primary or secondary amine, usually as the hydrochloride, with formaldehyde and a compound containing at least one hydrogen atom of pronounced reactivity. The essential feature of the reaction is the replacement of the active hydrogen atom by an aminomethyl or substituted aminomethyl group. The product from acetophenone, formaldehyde, and a secondary amine salt is an example.

The product from a methyl ketone contains reactive hydrogen atoms, and in some cases it is possible to carry the reaction one step further, yielding a compound with two basic groups.

Reaction Scheme:

Requirements:

Dimethyl amine Hydrochloride: 5.3g   paraformaldehyde: 2 g

Acetophenone: 5.9 ml   R-spirit: 8ml

Acetone: 5 ml

Procedure:

Place 5.3g of dry dimethylamine hydrochloride, 2.0 g of powdered paraformaldehyde and 5.9 ml (6.0 g) of acetophenone in a 50 ml round bottom flask attached to a reflux condenser. Introduce 8 ml of R-spirit to which 2-3 drops of conc.HCl have been added and reflux the mixture on a water bath for one_hour. The reaction mixture should ultimately be almost clear and homogenous. Filter the yellowish solution (if necessary) through a preheated Buchner funnel. Transfer the filtrate to a 100 ml conical flask. While still warm, add 40 ml of acetone. Allow to cool to room temperature and then cool in ice. Filter Off the crystal at the pump, wash with 2-3 ml of acetone and drain well.Crystalise the product by dissolving in 9 ml of tint R-spirit and slowly adding 45 ml of acetone.

Melting Point: The reported melting point is155-156°C.

Purification of organic compounds

A) Heating under reflex:-

Many organic reactions occur very slowly at room temperature. The rate of all reactions are increased by raising the temperature and a reasonably rapid rate can be achieved by carrying out the reaction at the boiling point of mixture of reactants or the solution of reactants if they dissolved in a solvent.

The apparatus for this purpose consists of water condensers attached vertically to round bottom flask. When the mixture in the flask is boiled, its vapour condenser at the cold surface of the condenser and liquid runs back into the flask. This is known as heating under reflex.

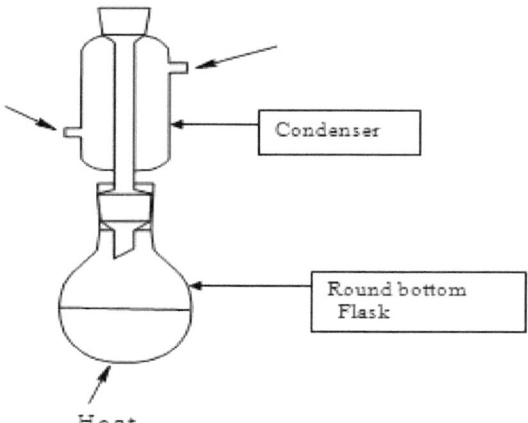

B) Distillation: -

When the product is a liquid or solid with a boiling point below about 250 ·c and no other volatile compounds are present, the simplest method of purification is by **distillation**. The water condenser is used for compounds boiling up to about 180.c and air condenser for compounds of higher boiling point. Mixture is heated to boiling and vapour of the product liquefies in the condenser, the liquid collecting in a receiving flask at the end of condenser. It is important to place thermometer so that its bulb is fully immersed in the stream of vapour which is about to enter the condenser.

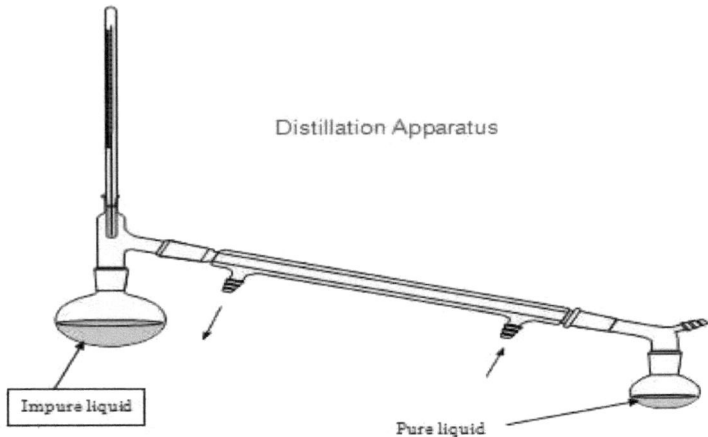

Distillation Apparatus

Impure liquid

Pure liquid

Some organic liquids decompose before they have a chance to boil at atmospheric pressure; they can still be purified by reducing the external pressure until their boiling points are below their decomposition tempeture. This is known as **vacuum distillation**.

When the required product is accompanied by one or more other compounds of similar boiling points, **fractional distillation** must be carried out.

If a liquid of higher B.P. is immiscible with water it can be purified by *distillation in steam*. This method is particularly useful when separation from inorganic and other non-volatile materials is necessary. The steam is passed through the heated mixture, and required liquid and water are collected in conical flask.

C) Extraction by solvent:-

If material to be purified is soluble in one solvent, while impurities are not, the mixture can be partitioned between two immiscible solvents. For e.g. sometimes an organic compound obtained as an aqueous solution, which is contaminated with inorganic materials. The aqueous Solution is shaken with ether in a separating funnel. Only the organic comp.is soluble in ether, the residue remains in aqueous Layer. Two layers are separated, ether solution is shaken in presence of a solid drying agent such as anhydrous magnesium sulphate to remove small amount of water which will have dissolved in it. The solution can then be filtered from the drying agent and org. compound can be separated from ether by distillation. Extraction is more efficient if several small quantities of ether are used, rather than one large quantity.

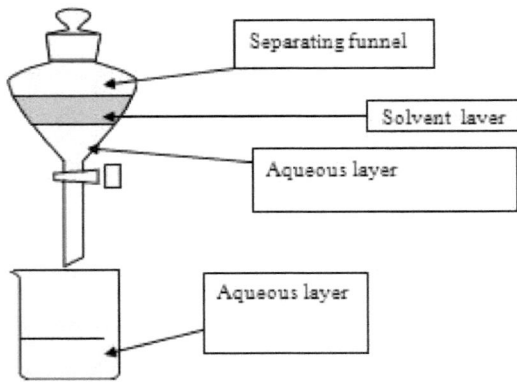

D) Recrystallition:-

This is the commonest method for purifying a solid. It is based on the fact that the solubility of organic compounds in a particular solvent increases as the temperature is raised. The solvent is heated with that amount of the solvent which gives a nearly saturated solution at the b.p. The solution must then be filtered very rapidly, and this is done using a fluted filter paper, contained in a glass funnel from which the stem has been removed, through which the filtrate runs into a conical flask; insoluble impurities are removed. The solution is then allowed to cool; crystals of the solid continue to be deposited until the solution has reached room temperature. The crystals are filtered off in a Buchner funnel, which is attached to   Buchner flask whose side-tube is connected to water-pump. When filtration is complete, the crystals are washed with a small quantity of the pure solvent to remove any impurities which might have deposited on the surface. The crystals are then dried on a watch-glass to remove the solvent. If one recrystallition does not yield a pure material, further recrystallition, preferably with different solvents, should be carried out.

E) Tests of purity:-

In most laboratories, the melting point of a solid substance and boiling point of a liquid substance is considered a sufficient indication of its purity.

Melting Point:-

A pure solid substance melts sharply at a definite temperature, while an impure substance will have a lower and indefinite melting point.

The crystals are powdered finely and charged into a capillary tube sealed at one end. The substance should stand in the capillary 3-4 mm from the bottom when thoroughly packed. The capillary tube is wetted with the liquid in the bath and placed alongside a thermometer fixed in stand. The capillary remains sticking to the thermometer by itself and is so adjusted that the solid in it stands just opposite to middle of mercury bulb. The beaker is heated slowly and temperature of bath kept uniform by gentle but constant stirring with a ring stirrer. When the substance capillary just shows signs of melting, the burner is removed and stirring continued. The temperature at which substance just melts and becomes transparent is recorded.

Boiling point:-

A pure organic liquid boils at a fixed temperature which is characteristic of that substance.

(1) Distillation method: if enough liquid is available, its boiling point is determined in an ordinary distillation apparatus. A pure liquid will distil out at a constant temperature which is its boiling point. In case the liquid is impure, the boiling point will rise during distillation.

(2) capillary-tube method:-when only small quantity of liquid is available, its B.P. is determined by the 'Capillary – tube Method'. A few drops of the liquid are placed in a thin-walled small test-tube .A capillary tube sealed at about one centimeters from one end, is dropped into it. The glass tube containing the liquid and capillary is then tied alongside a thermometer so that the liquid stands just near the bulb. The thermometer is then lowered in a beaker containing paraffin oil or sulphuric acid.